FERTILIZER FOR THE FUNNYBONE II

A COLLECTION OF SHORT STORIES AND ANCEDOTES

BY
GARY GUEST
(DOUBLE G)

Words Matter Publishing
P.O. Box 531
Salem, Il 62881
www.wordsmatterpublishing.com

ISBN 13: 978-1-949809-48-0

Library of Congress Catalog Card Number: 2019953604

"Sometimes in the ag world today, the ability to laugh at ourselves is the only thing that keeps us from crying. I hope that something I've written here will add to the total number of laughs you accumulate in your life."

Double G

Hello.

Welcome to Fertilizer For The Funnybone II. This is another collection of short stories and poems, and one that I hope will make you laugh or smile.

All of the stories are true and…most of the stories are…some of the stories…. At least one of the stories is true. Some of the names have been changed to protect the innocent. And some of the names are real because… well…they didn't need any protecting. The stories feature some of my childhood friends as well as family, including my wife, who not only serves as my on-site editor, but she is also an ~~un~~-willing participant in many of my tales. While putting together pieces for this book, I thought I might include a cooking recipe or two to show off my culinary skills, but I'm pretty sure how to cook a hot dog on a stick has already been covered.

Thank you one and all.
Gary Guest
(Double G)

Table of Contents

Acknowledgment

For this book, special thanks go out to Art-Tech of Nashville, Illinois for providing all of the illustrations that you see here. Art-Tech is my brother and sister-in-law, Don and Shelley Rybacki. Thank you guys so much. It is greatly appreciated.

Gary

MY REPLACEMENT

SOME OF YOU READERS MAY have come to believe that I can be a little goofy. I enjoy coming up with quips to try to get a laugh and often I'll break out in a song if I think it will draw out some smiles. Now I believe that a lot of my family is the same way, as whenever we're together laughter is a common occurrence. If there is such a thing as passing the torch, then I believe I may have found my replacement. While driving the other day, I had two of my grandsons as passengers, Briar is 11, and his little brother Kase is four. Briar and I were talking about being careful when putting on shoes or boots that haven't been worn in a while, because there might be something inside them.

He said one time his dad had put on a boot stored in the garage, and there were some wasps in it. I replied that when I was a teenager I remember keeping the boots that I wore in the hog barn outside by the front steps. One day while slipping them on, I felt something wiggling in my boot. In a panic, I kicked the boot off, sending it sailing away. As it did, a half-grown rat emerged from the boot. Without missing a beat, Kase joined in the conversation by breaking out in a song as he said, "Boy I bet that rat thought…*I believe I can fly*…. *I believe I can touch the sky*." His Comedic Timing Was Perfect. Bob Hope would have been proud of him. No more applications are needed. I believe the position of family comedian has been filled.

Double G

BEWARE OF THE SQUALL

As I sat at the kitchen table contemplating a writing subject, my wife walked by on her way to the laundry room, a.k.a. the basement, and casually brushed something off my shoulders.

"What was that?" I inquired.

"Oh, I was just brushing off some white hairs on your shirt."

White hairs? "Well," I replied, "they must be from the dog."

"Um, sweetie. We don't have a dog. And the white hairs are from your head."

"My hair is not white. It's just a very pale blonde."

"No. It's white. It has been for some time."

I rushed to the bathroom mirror, and somehow, the guy looking back at me did appear to be a tad older than I remember. As I checked my scalp, I realized my wife could be right. There may just be a few white hairs mixed in with my abnormally-pale blonde hair. Anyway, if there are any white hairs up there, I have a hunch that I know when they got their start.

I hate to tell you this, but I'm not an overly brave person. There have been things that have scared me all my life and continue to do so. Nowadays, things like health, weather, the economy, and my checking account balance all cause me concern. The other day a huge fear ran through my body when my granddaughter told me she was going out on a date. I have a granddaughter old enough to go on dates! Why…that would mean that I'm at least 30 years old! My wife tried to explain this phenomenon to me, but I'm afraid I wasn't listening.

I remember being a kid and discovering that scaring my two little sisters was something I had a knack for. I would leap out of my hiding places and growl or yell "boo" and smile as they ran screaming to mom. That's when my dad confessed to me that he was afraid also. I think his exact words were, "Son, I'm afraid that if you don't stop scaring your sisters, my hand and your posterior are about to have a meeting." Those may not have been his exact words, but their meaning was clear to me. I decided right then and there to retire from the scaring business.

As I said at the start of this story, I wasn't a particularly brave kid myself. I wasn't overly fond of the dark, and I tried to avoid scary movies. A friend might call and ask, "Hey, you want to go to the movies Friday night?"

I'd reply, "Sure! What's showin'?"

"It's Dracula versus Frankenstein."

"Ummm. I just remembered. I think I'm scheduled to be sick on Friday. Sorry."

Even though we hadn't studied health in school yet, my instincts told me that human blood was supposed to be inside the skin, not on the outside.

Flash forward to the summer of 1971. I had just received my driver's license, and I had my dad's permission to drive the farm truck. Actually, the farm truck was our only truck. It was a 1962 Ford half-ton with a four-speed manual transmission. Nowadays the technology is available that allows our vehicles to run on just a few cylinders until more power is required and then the rest of the engine kicks in. Our truck back then was ahead of its time. Even though the engine had six cylinders, if you had four of them firing at any given time you were lucky.

One hot August evening, my friend Gene called. He and another friend Scott were at his house, and he wanted me to come over for some fun.

"Um," I replied. "I'm not feeling real well. What'd you have in mind?"

It seems his sisters and some of their friends were planning a campout, and he had a great idea of how we could scare them.

"You know," I said." I'm suddenly feeling a lot better. I'll be right there."

When I arrived at his house, Gene told me about his scheme.

Gene's two sisters, two of their cousins and two more friends, were planning on camping out that evening. Their destination was a beautiful little spot about a mile from Gene's house and a quarter of a mile behind Gene's uncle's place. This was a place we were well familiar with since we camped there a lot ourselves. It was a wonderful campsite with a fire ring and tent situated in a stand of large trees in between two little lakes. So many of us kids camped there that Gene's uncle had purchased the tent, put it up and let it stay there all summer for whoever wanted to use it. The tent was 12 foot long by 8 foot wide with zip down windows and door. It even had a floor in it to help keep out the crawly critters.

The girls who planned on camping there that night ranged in age from 13 to 18. They drove to the site in two vehicles. Gene's dad's pickup truck and a red Volkswagen "beetle."

Gene's plan was for us guys to wait till it got good and dark and then sneak in near the tent and make noise to try and frighten the girls. I thought for a moment about what could possibly go wrong with this plan, but since my father's threats didn't say anything about my friend's sisters, I figured I was all right.

Most of the people going to these lakes used the road going past Gene's uncle's place, but there was a back road through the timber that we knew about. An hour or so after it got dark, we took my pickup and came in the back way. There was just enough moonlight so that we could drive without the headlights on till we got within a quarter of a mile from the lakes. We walked down a dirt road for a ways and then snuck through a standing bean field till we got within 50 yards of the tent.

We laid on our bellies as we peered through the last few rows of beans. The talking and laughter coming out of the tent was very loud.

I looked at my two companions and said, "As loud as those girls are, they won't even hear us when we growl."

Scott smiled and said, "Oh, they'll hear this." He proceeded to pull some kind of call out of his pocket.

I answered, "I'm sorry. I just don't think they're going to be all that scared of a quacking duck."

"It's not a duck call you dummy. It's a coon squaller."

"What's a coon squaller?"

"Well, it kind of imitates two raccoons fighting. When you blow it real hard, it makes a lot of noise."

We three guys smiled evilly as we thought about how scared the girls were going to be.

Scott said, "Here goes." And he sucked in as much air as he could.

When his lips hit that call, out came one of the most bloodcurdling, hair-raising sounds I had ever heard. As the echoes of that scream drifted off through the timber, I immediately became aware of the sounds of a large terrified beast crashing through the standing beans.... Then I realized...it was me.

I returned to my companions as quickly as I could and explained to them that just as Scott was about to blow on the squaller, I noticed a Sasquatch sneaking up behind us and so I fearlessly chased it away with little regard for my own well-being. In between their fits of mirth, they thanked me for my bravery. Then Gene crinkled up his nose and said, "Wow, Something stinks." He looked at me.

"Did you step in something?"

I then explained to them that it was a well-documented fact that Sasquatch sightings are often accompanied by foul odors. Thus adding validity to my story.

We then turned our attention back on the tent. We noticed that not only had the girls stopped their talking, but all the sounds of the forest had ceased as well. I imagined that the crickets had just rolled over and died from fear and all the frogs within earshot had swallowed their tongues at the same time. In fact, it was so quiet we could hear beads of perspiration striking the tent floor.

After about five minutes, we began to hear whispering going on inside the tent. I figured the older girls were telling the rest of them that the sound they heard was just a Tyrannosaurus Rex devouring a mountain lion. After another five minutes the volume of their talking was about back to normal. However, their laughter was being held to a minimum.

It was time for another blast on the squaller. This time I was more prepared when Scott belted out another scream. I had my fingers in my ears. Well, my fingers worked wonders on sparing my brain, but somehow my heart heard it and attempted to leave the area. It ricocheted around inside my rib cage, looking for an escape route. I kept my teeth tightly clenched so it couldn't go that way and the only other exit it could think of was a little too distasteful, so it decided to stay where it was.

Again, utter silence from the tent. Then we heard the zipper on the door come down a foot or so. Presently, a flashlight appeared with a hand attached to it. I know they meant for the powerful beam of light to sweep the bean field, but with the way it was dancing all over the place, a passerby might have assumed they were trying to flag in nearby aircraft.

We guys tried to curtail our laughter as we envisioned the frightened girls inside the tent. We didn't know how much longer they could hold out, but us guys decided we needed to end these festivities pretty soon. The reason was that we were starting to get lightheaded from the loss of blood. Us three laying out in a moist bean field and a hot August night was presenting the local mosquito population with a virtual smorgasbord. In the quiet of the night we could hear the little plinking sounds of mosquitoes exploding in midair from overeating.

When Scott blew the squaller the third time, the girls calmly left the tent and drove back to Gene's house. At least that's the way the girls told the story a couple of years later.

The way us guys remember it was, a mass of humanity with twelve legs exited the tent and flew in the truck cab in about a half a second. But then again we weren't watching the whole time because we had our heads down as pieces of the tent zipper ripped through the bean field like shrapnel. The fact that all six girls fit in the single cab pickup was also a bit of a surprise. The pickup was a four-speed with the stick on the floor, and I wondered how they managed to work the gears. When I asked the driver about that a few years later, she said she doesn't recall shifting in the sixty seconds it took them to get home.

After they left, us boys went and inspected the tent. I wasn't too surprised to see black marks on the floor of the tent from where they peeled out, but then I realized from the shoes that were present the girls must have been barefoot. Also, the black tire marks that the truck left on the dirt road managed to wash away in six or eight months.

We figured that the girls would head back to Gene's house and we thought it would be fun to hear their stories. We walked back to the truck. I opened the door and hopped behind the wheel. The other two opened the passenger door but then stopped and stared at me in the illumination of the truck dome light. Gene said, "When did you dye your hair white?"

I laughed as I looked into the mirror, "Are you crazy? My hair's not whi…. What I meant to say was, I dyed it a light blonde to see if girls might like it better."

"Looks white to me."

"Well, it's not. It's light blonde. Now get in."

We drove to Gene's house and pulled up close to the yard gate. Gene's parents were out in the yard, and they were listening as all six girls tried to tell them about the hideous monster at the lake.

At 16, I had not yet faced many life or death situations, but something in my natural survival skills told me to keep the truck running. Scott, who was seated by the passenger door, crawled halfway out and sat with his butt on the windowsill so he could look across the roof of the cab. When all of the girls happen to take a breath at the same time, Scott let out a little blast on the coon squaller. At first, six sets of eyes went wide open. Gene's parents figured out what happened and fell to the grass laughing. The wide-open eyes all narrowed down into little slits and fire started emanating out of them. I don't recall my brain telling me to get out of there, so I guess my feet just acted on their own. One came off the clutch while the other mashed the accelerator. By the time we were a mile down the road the girls started to fall further behind. If they would have been in a vehicle, they would have undoubtedly caught us. That or if they had been wearing shoes. I told the other two that I thought we should swing by my place and pick up a few provisions and maybe hide out in the woods for a while. I asked Gene, "When do you think it will be safe to go back?"

He replied, "They should cool down in three or four… years."

We stayed away six, just to be sure.

Double G

A CATTLEMEN'S TALE

The raccoon's name was Bandit, and he was
true to his name

Robbing and stealing were a part of his game

He raided the henhouse each night about two
And he knew to be careful, don't wake up ol Blue
For Blue was a coonhound that guarded the place
And if he caught wind of Bandit, there'd be
quite a chase
These two had matched wits on many a night
And on occasion, they had them a fight
But somehow the bandit would just get away

The whole thing was like some game that they'd play
But bound and determined was this coonhound
named Blue
He had him a plan, and he thought it all through
He'd sleep all day long and when came the dark
He'd be at the ready and by the henhouse he'd park
When Bandit came calling he'd have a surprise
This might be the night of the old coon's demise
Then old Bandit came sneaking just like before
And jimmied the latch and opened the door
Well ol Blue was ready and he came on the run
Bandit reared up and the battle begun
The fighting was fierce but short in its length
Ol Blue had won out, he just had more strength
There's a lesson here folks and it's easy to see
If there's wings on a menu wherever ye be
Know this craving for poultry might add to your grief
The moral of it all, is to just eat more beef

Double G

ME, STUBBORN?

ON OCCASION, I HAVE HAD a friend or family member say that I am a tad bit obstinate. It's just that I have no intentions of coming around to their way of thinking when I truly believe that I am correct. Okay, so maybe that is the definition of being stubborn. I do know I had an ancestor named Francis, but I'm pretty sure he wasn't the mule that starred in the movies with Donald O'Connor. I'm certain that my buck teeth, large muzzle, and long pointy ears are from some other ancestor.

While sitting here and writing this story, my wife looked over my shoulder, "What're you writing about this time, sweetie?"

"I was just explaining to my readers that I am not a stubborn guy."

I really wish my wife wouldn't drink that orange soda. It looks so disgusting when she laughs so hard that it shoots out her nose like that.

Okay. So maybe I have a little bit of a stubborn streak in me. Growing up at home, we used to tease my mom about being the stubborn one of the family. And maybe she was. I know that my two sisters and I can all be a little opinionated at times. My brother, however, is not a bit stub…. Oh wait a minute, I was thinking of somebody else's brother. Yeah, my brother's stubborn too.

I would make the case that being stubborn about some things is not all bad.

Nowadays there are all kind of ideas going on that the world says is okay. I kind of think the values that we grew up with might be a little better. The most important board that we had at our school was the one that hung in the principal's office. And he used it. Not that I would know firsthand. But that's what I was told.

And I would be okay with it if the nativity scene was erected each year on the courthouse lawn like it used to be. The money we use still says, "In God We Trust." Seems like we're telling our maker, it's okay to be involved in our money but don't be messing with the decisions our judges make. Yeah, I kind of wish He was. Gee. The way this story is going, it sounds like it's being written by a crotchety old man.

I just went and looked in the mirror.

Yikes! It _is_ being written by a crotchety old man!

One area where I tend to drag my feet a little bit is in the world of technology. I'm kind of comfortable in the old ways. Oh, I know that the world is constantly changing and guys like me shouldn't stand in the way of progress. Seems like every day there is some kind of electrical gadget coming out to make the world ~~lazier~~ easier. My grandkids are amazed that I have a flip phone, and it still works. Kinda. Sometimes you have to bang it on the table a time or two or maybe call it a bad word, but it still works.

I guess the changes that have come to our tractor driving bothers me the most. With guidance, auto steer, and driverless tractors, I'm not certain that you're allowed to identify yourself as a farmer. Growing up, we took some pride in learning how to drive a tractor and do it well. One of the rites of passage was proving to my father that I could mowboard plow straight enough to be allowed to plow out the furrows. Now, as some of the younger readers peruse this article, they look at each other and say, "What's a mowboard plow?"

Another area where we worked hard was planting. You wanted to impress the passersby with neat, straight rows. The neighbors who couldn't plant straight came up with sayings like, "You get more seeds in a crooked row." I can't prove it, but it's quite possible that it's the offspring of these guys that invented guidance and auto-steer. With the help of these inventions, now almost everybody's rows are straight. A couple could be driving down the road today, seeing thousands of acres of arrow straight rows, then, they pass by my field.

"Well, looky there, Alice. It appears that somebody tried to plant that field while steering the tractor himself!"

"The poor man!"

Actually, if you take money out of the equation, I've always considered myself to be pretty rich.

Nowadays, I find myself turning more and more of the farm tasks over to my brother, my sons, my son-in-law's, my nephews and my dog. And the farm is definitely running better. Don't you hate it when the dog gloats?

For years and years, I was the main guy at applying all the anhydrous ammonia. Now my one son does most of that. He has the tractor set up with guidance and auto-steer and most every other invention that has come along. This past spring, he had to leave early one day, so he called me and asked if I could take over. I said I would. When I sat down in the seat he began to give me a lesson about what each control box did. I really wasn't listening that close, because whenever he left I had every intention of shutting all these contraptions off. I waited till his truck disappeared, then reached for the main power switch. A voice came out of the control panel, "What do you think you're doing?"

Somewhat startled, I replied, "Who said that?"

"I am the master controller of this tractor. I am **Field – Ready – Excellent – Driver**. But you can call me Fred. I was named after my inventor."

Still amazed, I replied, "Say. I remember an old guy down the road named Fred. Boy, were his rows crooked."

"That was my grandfather."

"So you understand everything I'm saying?"

"Absolutely, Gary. I can speak and comprehend fifty different languages."

"Wow. Say, how'd you know my name was Gary?"

"Once you touched the door handle, your fingerprints were analyzed, and your name was produced from my databank. So why were you going to turn me off?"

"Well. I guess I was just wanting to experience tractor driving the way it used to be. Kind of reminiscing, I suppose."

"So you were going to shut me down. What would I do when I went home to my little transistors and told them daddy lost his job today?"

"Gee Fred, I'm sorry. I didn't know. Look, I'll just go home, and you can go back to work."

"No, no. Please stay. Now that we understand each other, I would really appreciate the company. Besides, if I go home too early, the wife will just start complaining about how bad the kids were and how tough her day was and then she'll say that I'm not paying attention and the next thing you know she'll blow a fuse. You know how it is."

"You're preaching to the choir, brother."

"You just sit back and relax. If you look in that compartment to your right, you'll find some cold beverages that you might enjoy."

"I'm not sure I should drink and drive."

"You're not driving."

"Good point. Say, you got a bottle opener handy?"

"Next drawer down."

"Thanks."

Fred was turning out to be a really nice guy…uh machine. Still, my mind was whirling with the whole idea of talking to a computer. I snapped back to attention when I detected a very pleasant aroma. "Say Fred, are you overheating?"

"No. That would be your supper. I just got it done. Push that yellow button that just lit up on the right side of the console."

I pushed the button and out popped a plate of food that would put most five-star restaurants to shame. I was hungry, and it didn't take long before my plate was empty.

"Well," Fred said anxiously, "How was it?"

"Oh man Fred. It was fantastic!" I didn't want to hurt Fred's feelings, but the filet mignon was a bit undercooked for my liking. And the fresh asparagus spears were cut into pieces smaller than I prefer. But the au gratin potatoes and the baked Alaska were just perfect. We rode around for another hour or so telling stories and bad jokes. And then Fred announced the field was done and it was time to go home. I stood up in the cab and began to trace the main wires leading into the back of the computer.

"Uh, Gary…you're kinda freaking me out. What are you doing?"

"I'm trying to figure out how to take you with me. You see, I drank the rest of those beverages, and you're going to have to drive me home. If I drive myself home it's going to be <u>my</u> wife who blows a fuse."

Fred replied, "Pull the red wire first."

Double G

FRIEND BUCKY

I'M ONE LUCKY MAN, I know I've been blessed
Cause life can be hard, it can be quite a test

And to live free and happy we know can be a big chore

There is labor to do and more bills to pay

As we steward this land and strive day by day

To give it our best and see what fate has in store

No, farm life's not easy, we always know that

Each morning I rise and I put on my hat

And head on outside to face the challenges there

Some days the burdens can be quite a weight

You clock in real early and clock out real late

But to face it alone, well that might be too much to bear

That's when I'm grateful that help is around

There's family and friends to take care of ground

And to get the chores done no matter the time it may take

We all work together to get it done right

Even if the labor spills into the night

And will do it again when tomorrow's sun doth us wake

I could go on for hours about the people around here

Bout the neighborhood folk and the family so dear

But the words I would use would somehow not be enough

They can pick up my spirits as we battle a task

And to see it all through is all I can ask

And without these people, this life just might be too tough

But before I should end with this bit of prose

There's another to mention before I do close

And this one is special in a way I hope you'll agree

Because my constant companion all over the farm

Is a four-legged kind with a heart that's as warm

As the rays of the sun as he walks right beside me

Bucky is there, come rain or come shine

He doesn't complain, you won't hear him whine

Being my sidekick is all it takes to be happy and glad

He listens intently if I start to wail

And when I glanced his way, he just wags his tail

And just seems to say, "Now Gary, it can't be that bad."

I'll let out a smile and give him a pat

And thank him for being right here where he's at

And for helping to put my nerves back on the mend

I bet as you read this, you'll think of a pet

That's a special a companion as you'll ever get

And you'll understand when I say…"Bucky's my friend."

Double G

I MUST BE ON THE A-LIST

I WISH MY COWS WERE AS good at reproducing as my wife's "to do" list. It seems that if I do manage to complete one of the tasks written there, the list gives birth to three more chores almost instantly. I admit that I am not very good at completing the chores that she has designated for me, but in my defense I say that I am a really busy guy. I mean really. If I didn't frequent the coffee shops, the feed stores and the machinery dealerships in the area, where would all the other farmers get their information? I'm sure that the farm wives wouldn't want their husbands picking up ~~gossip~~ news from just anybody.

Every so often I'll hear of some guy who actually claims to have accomplished all of the chores on his wife's list. I tried to avoid guys like that because for one; they make me look bad and for two; I don't want to become as big a liar as they are.

I can't prove it, but I'm pretty sure there are some things on my wife's "Honey-do" list that were written there before she met me. Talk about foresight.

Every once in a while I'll take the time to get a few of these chores done. Like, trim the hedge, clean the gutters, take the car to the vet, wash and wax the cat. In hindsight, I wish I had read through those last two chores a little more carefully.

I think that lists have always made me feel a little uneasy. I'm sure it started with Santa Claus's naughty list. In the days of my youth, my parents were constantly reminding me that bad little boys that are on Santa's naughty list won't be getting any presents. Now, I remembered getting presents the year before, so I imagined I must have received a stay of execution right before the big day last year. But being good this year was proving to be more difficult. And, adding to my troubled young psyche was the fact that Santa must be getting tipped off about my deeds by my parents. Apparently, they had a direct phone line to the North Pole. Years later I discovered more evidence that my parents and Santa were in cahoots when I found all my alibi letters that I had written and that my mother had promised to mail for me. One year my father sat me down and said, "Son. You really need to try to get off the naughty list. I mean really, son. You're twenty-one years old!"

Actually, I thought that being twenty-one was a great reason to be on the naughty list. But I reasoned that was another conversation for another time. So I promised dad that I'd be good and I tore up the alibi letter that I had just written.

Shortly after that, I realized my foot speed wasn't as good as it used to be and my future wife caught me, and we got married. That's when the naughty list was replaced by the honey-do list. It's been with me ever since.

At this point, as I was writing this story, my wife walked through the room, and I said, "Hey honey. I'm writing a story about lists. Remember that time back in eighty-nine when we worked our way onto that wedding list at that hotel we were staying at? You remember we left the two kids with your parents for a couple of days."

"Oh, I remember," she replied. "And it was four kids."

"Four kids?"

"Yes. By eighty- nine, we had four kids."

"I thought those other two were just visiting."

"No, they were yours too."

"Gee, you would have thought I'd have remembered that."

"Yah, you would have thought."

Well, regardless of the number of kids, the incident to which I referred, happened like this. At that time, my wife and I had reached the pinnacle of our economic status. By that I mean we were merely broke. We went on a weekend vacation to the city. While we had budgeted the necessary funds for the hotel room, it left little cash for food. When I would go across the street, the kid waiting on me at McDonald's started looking at me funny when I would order to happy meals with coffees. In our hotel room on that Saturday night, I began to hear the sounds of a big party going on below us. I went downstairs and checked it out. It was obviously a wedding celebration, and through the open doors I spied tables full of food available to the invited guests. I went back upstairs and told my wife to put on her best dress. We were going to crash the party.

My wife replied, "Oh, really, honey. I don't think we should do that."

"Relax," I said. "They have way more food than they can possibly eat. We'll be doing them a favor."

Against my wife's better judgment, she got ready, and we went downstairs. We walked through the open doors and were immediately met by a large bouncer with a clipboard.

"Name, please?" he inquired.

Somewhat startled, my mind raced, searching for a name.

"Heinrich," I said.

He looked down his list. "There's no Heinrich on the invitation list."

"Oh, we must be on there. Me and the groom's father are old pals."

"Sorry. You're not on the list."

I continued, "You mean to tell me that everyone who was invited is already here?"

He looked at the list again. "Everyone has arrived except the Ortegas."

"That's us! I'm Heinrich Ortega. My mother was German, my father was Spanish, so they compromised on my name."

Then the bouncer made a mistake. He blinked. When his eyelids came back open, we were in the middle of the crowded room. Either he bought my story, or he figured we weren't worth the trouble to look for. I guess we blended in pretty well although we were the only couple wearing a square dancing dress, cowboy boots, jeans and a bolo tie. Everyone there was very friendly, and the food was fantastic. By the time we left we were so full we could barely make it up the stairs to our room. I bet the kid at McDonald's was wondering where I was at the next morning.

That all happened a long time ago, but lists are still a part of my life. As I was wrapping up this story, my wife came up behind me and put her arms around me and gave me a hug. She kissed me on the cheek and softly said, "I have another list. It's a list of who I would like to spend the rest of my life with and there's only one name on it." She smiled and walked away.

Boy, I'd really like to get a peek at that list and see who it is.

Double G

AHHHH...CHOO
WINTER

WINTER TIMES HERE AND THE earth is aglow

The hills are all shining with the new-fallen snow
The temperature's cool and our cheeks are all cherry
As we gather together and say, "Have a merry…"
The children all giggle as they slide down the hill
With their eyes opened wide, oh man what a thrill
There's a fire brightly burning and there's cocoa
for some
Or maybe some brandy or a quick shot of rum

It's really quite pretty, but there is just this one thing
As I tend to my cows, I can't wait for spring
I love all my cows, I have all my life
I think like a cow or so says my wife
Watching them graze in the meadow is grand
The animals are truly a part of this land
But when wintertime comes it would be so great
If I could teach them to just hibernate
The wind is whipping, it's a mite hard to see
The flurries are swirling and some stick to the tree
The temperatures cold and there are patches of snow
And the people all scurry wherever they go
Their vision is skewed with the fog on their glasses
And if they should slip, they'll fall on their sidewalks

NOT TONIGHT HONEY…I'M TOO TIRED

WELL, I THINK IT'S TIME I address a subject that many other writers are afraid to talk about. If there are those around you that are offended by the use of certain language, you may wish to ask them to leave. I mean really…. Some people just can't be rational when you're talking about…tires.

The caveman named Ock that invented the wheel was a genius. He took a rock and chiseled it to a round shape.

He placed these round rocks on his trailers and wagons thusly making life much easier for him and his friends. The argument could be made that Ock was much wiser than today's inventors because he didn't try to put a rubber donut around his wheel and then fill the donut with air. I can't prove it, but I wonder if the tire wasn't invented by the guy who first invented the bumper jack. I mean what good is a bumper jack if you don't have flat tires.

The story I was told was some guy named Retread invented the tire. He worked night and day promoting his invention. One night when he sat down at the dinner table, he told his wife, "You know Akron,… I've had a really good-year." That's when Akron told him she was running away with a guy named Ock. This news left Retread totally deflated. A lot of men and their tires have been deflated ever since.

Nowadays, everything that moves has tires on it and society couldn't function without those rings of rubber. I suppose that my view of tires is a little irrational and I blame my early years on the farm for my opinions.

Growing up, I had heard a rumor that there was such a thing as new tires. I consider this to be just a rumor because all I ever saw was used tires. All our neighbors ran used tires as well. Now I'm not totally ignorant, regardless of what my teachers may have said, I understand that in order for there to be used tires someone first had to buy them as new tires. I'm not sure who that guy was, but I never met him. That's why me and my neighbors were always rooting through the used tire pile at the local service station in search of a tire that still had some life in it. Now the guys at the service station helped us tire scavengers out by "grading" the used tires. There were only two grades. O. K. meant…OK. And N.F.G. meant don't take it home.

One day I asked my dad what N.F.G. stood for. After first making sure mom wasn't around, he told me what it meant. This led to an uncomfortable situation one day when I was with my grandmother, and we stopped at the service station to buy some gas. While standing there, grandma noticed some tires with N.F.G. written on them. She asked the attendant what it meant. Now as the attendant stared at this 70-year-old Sunday school superintendent, his face began to go into convulsions as he tried to think up a suitable lie. At last he said, "It means No, Further, Go."

I laughed out loud and said, "No, it doesn't. It means…." My grandma looked intently at me. Then I continued, "Oh, yah…that is what it means."

Finding a good used tire was simply a matter of economics. On a grain and livestock farm back then it wasn't just your car, truck and tractors that had tires on them. You had oodles of old hay wagons, rakes, balers, and other machines in need of tires as well. You just couldn't afford to put new tires on all these things. That's why if we heard of a rich city person getting a new set of tires and only three of their tires were actually worn out, we would be standing by the used tire pile waiting for that good one. My philosophy on tires back then was shaped by my father. He used to say, "a tire isn't totally worn out until you can see the air on the inside of the tire." Believe me, we used this philosophy for years. It wasn't uncommon for a hay wagon back then to have four different size tires on it. And as you can imagine, anyone who ran as many used tires as we did had to be good at changing flat tires. I could do that in my sleep. And apparently, I did a few times. I'd come downstairs for breakfast and dad would ask, "Somebody jacked up the car and took one of the tires off last night. Was it you?"

I'd scoff, then reply, "No. Of course not." Then I'd reach in the pocket of my pajama pants and find a handful of lug nuts. I changed a lot of tires back then.

Adding air to an old dry, cracked tire was a very difficult job. The difficulty arose in trying to find somebody naïve enough to kneel down next to the tire and put the air in it. The rest of us would stand inside the nearest building with our fingers in our ears.

Back then, when I would drive the old pickup to school, some rich kid would come over and ask, "Say, Gary. I was just wondering what brand of tires you favor on your vehicles?" He would then walk around my truck while saying, "I see you are a Firestone, B. F. Goodrich, Goodyear, Uniroyal kind of man."

I'd reply, "Yes. Yes, I am."

We still run a lot of used tires on our older machines yet today. Most tires now are steel-belted radials. This brings a new problem for used tire users. Broken belts. The belts are steel bands that are molded into the rubber of the tire. When a band breaks, the tire will develop a lump somewhere on the tire. When you drive a vehicle that has a tire with a broken belt you will feel a distinct wobble to your car or truck. If you are a person who has never driven a vehicle with a broken belt but want to know what it feels like, you can duplicate that feeling in your home. Take an old quilt or comforter and get it soaking wet. Then put it in your washing machine. Make sure the wet quilt is all wadded up on one side of the washing machine so that the washer runs out of balance. Turn the washer on the spin cycle and then sit on top of the machine and imagine you have the steering wheel in your hands. Yah. That's kinda what it feels like.

Times have changed, and the demand for used tires has pretty well dried up. There's not that many livestock farms anymore, and the big tractors and large tillage tools require new or almost new tires to support them. I no longer spend time digging through tire piles looking for a prize. But every so often I'll be at the tire store, and I'll see a customer having a new set of tires put on his vehicle, and I'll see the tires he's getting rid of and think, "Oh my goodness! These things are like new!" Fifty years ago a set of tires like that would have caused quite a fight at the tire pile.

My wife told me the other day that she was glad I didn't go looking for tires like that anymore. She said, "Besides, you've developed quite a spare tire on your own."

Wives are so hilarious.

I think I'll head to town. I heard there is a new business opening up. It's called "Ock's Wheel Store." You never know. I may want to invest some money in this place.

Double G

THEO
THE THREE-POINTER

THERE WAS A DEER THAT my family knew
And behind our farm, this whitetail grew
And in his early years, he wouldn't drop your jaw
He was kind of spindly and kinda small
He hardly had any rack at all
And three little points was all our family saw
Theo the three-pointer became his name
And being quite puny was the source of his fame
And we hunters all laughed whenever he'd passed by

Now seasons come and seasons go
And Theo, he continued to grow
And in three years he began to catch our eye
Where once his headgear wasn't great
Now set a rack with points of eight
And this whitetail buck became a special stag
Now to all of us, it was plain to see
That on our walls we'd wished he'd be
And then to others, we could really brag
I said, "I'll take him, if I can."
And so I then devised a plan
To be the hunter that would claim this prize
I'd hunt real hard and hunt real true
And before next seasons through
I'd hold the horns that had grown so big in size
So that next year while in my stand
I saw him coming so big and grand
And I could picture Theo hanging on my wall
I smiled a bit and thought, "I win."
But then "Buck Fever" did set in
And I couldn't hold him in my sights at all
I began to panic and to fear
I wouldn't get my trophy deer
If I didn't stop my shaking, here and now
So I held my breath and closed one eye
I knew the time was now to try
And told myself, make sure some lead allow
I touched the trigger, the gun went bang
And the volume of the echoes rang
Throughout the families woodland far and near
When the smoke died down and I could see
I hoped there'd be a prize for me
But Theo seemed to up and disappear

I walked to where the whitetail stood
And there I found amidst the wood
A piece of Theo's horn there in the snow
I didn't hit him in the heart
Instead, I shot his horn apart
And this three-point rack was all I had to show
I put this horn up on my wall
And I'm quick to tell my tale to all
That day I decided to spare the old deer's life
I expertly shot his horn that day
And let old Theo run away
And they all, believe me, even my own wife
I wish the doctors of this great land
Would set to work and get at hand
To alleviate Buck Fever with a cure
And so, for now, I'll just brag
About the horn that I did bag
But only me and Theo know for sure

THE MacGYVER
AWARDS

DO SOME OF YOU GUYS and gals to do your own repair work? We do. At least as much as possible. Obviously, some repairs are best left to the professionals. Sometimes the choice to do the work yourself or not comes down to where you are. Does the repair require a shop, can you get the machine to the shop, and is your shop big enough for the machine. Our shop isn't real big, so a lot of our repairs take place in another shed are under a shade tree.

I get jealous when I watch TV shows or read in a

magazine about some of the shops being built nowadays. They had one on TV the other night, and although I didn't get the dimensions, it appeared to be about the size of Cleveland. If you were working on one end and needed a wrench or a part from the other end of the building, you had to take a four-wheeler or an ATV to drive to that end to get it. If you would have walked the length of the building, by the time you got back it would be quitting time. A friend of mine who saw the show said that the two ends of that shop were in different zip codes. I think my friend might have been stretching the truth just a little bit. Good thing I never do that.

Many of the shops out there have wash bays for cleaning equipment. They get more elaborate all the time as well. I saw one the other day that had a sunken area for pre-soaking their machinery. That or there might have been a drain that was clogged up.

No, our shop is not at all that fancy. But we do have most of the things that we need. Since we are a smaller farm, we try to save money where we can. (You can substitute the word "cheap" in here.) Sometimes we invent parts or tools to get a job done. Kind of like MacGyver used to do on television. You remember Richard Dean Anderson played this role where he almost always had to make something out of nothing in order to save the day. I think he could have built Noah's Ark using a box of toothpicks.

One night while my wife and I were watching the show, MacGyver and his friends got trapped in a cave by a rockslide. One friend said, "I can't find anything to help us get out." A second friend said, "Hey MacGyver, I just found this cutting torch, and this 8-inch long piece of baling wire." MacGyver said, "Stand back." The next thing you see is the Sherman tank that MacGyver built come busting through the pile of rocks.

I let out a laugh and told my wife, "Oh, come on! You know that's not real!"

My wife, who was sitting on the edge of her seat, munching her popcorn, turned towards me and scowled, "You're talking about my Richard. Besides, I don't say anything to you when you're watching "Baywatch" with that other Anderson on it…. They're not real either you know."

I took my bowl of popcorn and headed to my office to sulk.

My wife yelled at me, "While you're in your 'office' you can take the towels out of the dryer and fold them."

Yah, right after I build me a new shop, I'm going to build me a real office.

As I said, sometimes on our farm, we build the things we need. My brother and I rate our creations and if we think they're good enough, will give ourselves a MacGyver Award.

Most of the things we create are small. But a couple of years ago, I built a log grapple out of a mowboard plow. The plow was just for parts anyway so I saved the three-point hitch and took two of the arms of the plow that hold the mowboards. I took the mowboard's off, and that left me with two pieces of metal that resembled the letter J. I mounted the plow arms to the three-point hitch with a single pivot bolt and about two foot of the arms sticking up in the air above the pivot. I fashioned a hydraulic cylinder between the tops of the arms. When the cylinder is contracted the J's at the lower end of the arms close together and clamp onto the log. You can raise the three-point arms to help drag the log. Ta-dah! I got a MacGyver Award for that thing.

Usually, we create things because we are unwilling to

pay the price that the new item would cost.

When a universal joint broke on the steering shaft of our one tractor, the tractor company said we had to change both the shaft and the knuckle at a cost of over eleven hundred dollars. At a car parts store, I found a knuckle from a pickup truck and fashioned it in place at a cost of nine bucks.

While changing blades on our Turbo-chopper, we needed a bigger wrench than what we had available to hold the big nut on the other end of the arbor bolt. My brother cut the size of the nut out of the one end of a discarded rotary mower blade. It worked great. More MacGyver's.

We might think we are resourceful. You might think we are cheap. You're probably right. You could add to that I tend to procrastinate as well.

One time I was combining wheat with my 6600 John Deere. An ear broke off of the tightener spring on the chopper belt. This happened in the year 1986 B.C. The B.C. stands for before cell phones. I was several miles from home and only had a few more acres to harvest. A search through my truck turned up a small chain boomer. I got one hook onto the broken spring and the other hook to the combine frame. When I snapped the boomer tight, the chopper belt got snug enough to keep running. I thought to myself, if this just gets this field done, I'll get a new spring tomorrow. Four years later, when I traded off the combine, the boomer was still on there.

Another product we turn out at the shop is deer blinds. Now, our family enjoys deer hunting but being a rather large person, I don't have a great deal of faith in the store-bought stands that have weight restrictions on them. On the one model a family member purchased, I believe the upright supports were made of the same fiberglass that they used for making fishing poles. I prefer to build my own. We use two-inch square tubing for the uprights and the steps and used mesh hog flooring for the platform. When we put these blinds up in the timber, they help hold the tree up. Another bonus is if we have one of these homemade blinds at home we can lean it against a shed and hook a chain hoist to it to lift a motor out of a vehicle. They're strong.

We've come up with many other ideas over the years as the need arose. I'm sure many of you have created bigger and better things than we have. Maybe you should give yourself a MacGyver Award. I always wondered if MacGyver would give me a call and get my opinion about something he was fabricating…. He never called.

But, I did get a call the other day that momentarily raised my ego. My phone rang, and the woman's voice on the other end said she was Pamela Anderson and she was wondering if I would be willing to star with her in an upcoming episode of Baywatch? Then, in a sexy and sultry voice she added, "I would really show you my appreciation."

Well, I couldn't let the poor girl be heartbroken, so I said, "Yes!"

Then the voice on the other end burst into hysterical laughter.

You would think my wife would have better things to do than make prank calls like that.

Double G

MY NEW TRUCK

I HAVE FINALLY DECIDED TO GET a new truck
The old one I've got has ran out of luck
It's rusty and dented, and its miles number many
It sags quite a bit, and for spark there's not any
But if that truck looked at me, it probably say
"Hold on there now Gary, you look the same way."
So I went to the store to see what they had
The man clapped his hands, and he said he was glad
That I had stopped by and this was my day
You can pick out a truck and just drive away
Just sign on this line, that's all there is to it
I said there "Hold on, I'd like to read through it."

There's some really big numbers, and I had to look twice
And I swallowed my gum when I saw the list price
So I said to the guy, "This price listed here
Is more than I'll make after working five years."
He smiled, and he said, "Oh that's quite okay.
Well maybe a used truck would suit you today."
I said, "I agree, just what have you got?
And he took me outside to the used parking lot
The used trucks were nice, it was all plain to see
And I picked one right out that looked good to me
I said, "What's the cost of this set of wheels?"
He said to me, "Friend there all such good deals.
For the one that you picked, I need just this much."
I put my hand to my chest and to my heart I did clutch
Why that's almost as much as the truck that was new
I'm afraid on my budget, it's overpriced too
He said, "That's alright, I guess I can bend.
So tell me how much were you wanting to spend?"
So I told him a sum that won't cause me fear
He crinkled his nose, and he started to sneer
For that kind of money, I'll have to look more
I may have you something in back of the store
He came back and said, "It's the best I can do.
I don't often get a big spender like you."
So I've got me some wheels, but I'm not sure I like
The way that I look on my new ten-speed bike.

SLIP SLIDING AWAY

"**N**ow, don't forget," my wife said. "We have to be there by six tonight."

"I remember," I replied. "Um. Where was it again that we were going?"

"To the granddaughters birthday party!" My wife shakes her head, then adds, "Boy. You are really slipping."

After she was out of hearing range, I said, "I am not slipping. I just wasn't paying attention when you told me before."

I suppose I am slipping in some matters of the mind, but some of my most spectacular slipping occurs when

I'm just trying to do chores. You'd think with the extra weight I've put on over the years that my traction would have improved. But actually, the opposite is true. I'm constantly searching for a good foothold while performing a task. When the ground is muddy or icy, then get your video cameras ready.

I don't mean to demoralize our young athletes who are preparing for the Olympics, but some of the moves that I've involuntarily put on would make the figure skaters green with envy. Just the other day I put together a flip, barrel-roll, somersault, split and a half-pike all at one time. And I didn't spill the two five-gallon buckets of feed. I remind you that these kind of moves are not meant for amateurs. As I understand it, a triple-axle is a coveted move for figure skaters. That's nothing. During a fall the other day, I bounced off four sets of axles before coming to a stop under the feed tank. When I came to, my son was yelling how somebody had put a dent in his truck fender. He said the dent was about the size of a man's face. "In fact," he added, "there's a little indentation where a nose might have been."

Well, that may be where a nose is supposed to be, but at that moment as I felt around on my face, my nose was nowhere to be found.

Obviously, that kind of slipping and sliding is unintentional. But sometimes we actually want to lose contact with the ground, snow, or water as the case may be. Sledding along on a cover of snow is something our whole family enjoys. I enjoyed it a whole lot more before they invented arthritis.

My participation in a snow event these days usually consist of me shoving a grandchild's sled down a gentle slope or maybe driving an ATV and pulling a saucer around in a field. As a grandparent, we don't want anything that's too dangerous. But when we were kids, the danger factor seldom entered into our thought process. Me and my friends could have climbed Pike's Peak and not given a thought about sliding down that slope. Here's a bit of knowledge that current kids aren't aware of. Car hoods were actually invented for us kids to use as sleds. Apparently, one day somebody put their car hood over an engine compartment to let it dry, and a car manufacturer saw it and decided to make all their cars with engine covers. But before that, they were just sleds. The cool thing about car hoods was they were big enough to carry more than one person at a time. The bad thing about car hoods was that after you rode them down the hill, you had to drag them back up the hill. Sometimes, halfway back up the hill, you would turn around and look to see if an elephant was hitching a ride. For this reason, we mostly just pulled car hoods around behind pickup trucks or tractors. One time my friend Gene and I convinced his dad to drive the pickup and tow us up and down the old dirt field road behind their house. One thing that we had learned was that by throwing our weight from one side of the sled to the other, we could make the sled go from ditch to ditch increasing the thrill factor in our ride. The only time we had to be sure to be in the center of the road was when we crossed a concrete culvert. Gene's dad, driving the truck, had no reason to believe we'd forget this. We forgot. With the snow flying up in our faces we momentarily lost track of where we were on the road. We missed the culvert and became airborne as our sled headed for the far bank of the drainage ditch. When the sled reconnected with mother Earth, Gene and I were propelled straight up as if we had been shot out of ejector seats. I don't know how high we went or how long we were elevated, but it's safe to say we recorded more airtime than the first manned space missions launched from Cape Canaveral. It's possible we went higher than they did also. The main problem with being in the air that long is, it gives you time to think about what's going to happen when the law of gravity kicks in. Since we were only thirteen at the time, seeing our life pass before us didn't take long. So I started to think about the future…. Presuming there was going to be one. If mother nature was trying to do us in that day, she failed to take into account the softness factor of the foot deep snow. Our hitting the ground triggered the only known avalanche to have ever occurred on flat land. By the time we came to a stop, the better part of a ten-acre field had been cleared of snow. Gene's dad hauled us and the sled back up to the house. Upon examination, numerous dents were discovered…some of which were on the sled. That ride was probably the highlight or lowlight of my sledding career. I only wish someone would have captured it on video. But not the audio part. If that had been recorded it could have been used as training in Marine corp. boot camps in regard to improper language.

Slipping and sliding on snow or ice can pretty much be achieved any time the temperature is cold enough. But skimming across the surface of water takes another ingredient. Momentum. If you're going to go water skiing or tubing, you need a boat to provide you with adequate speed to stay on the surface. I've always enjoyed watching people ski. Most of my family can do it. Me…not so much. In order to get me out of the water, it would take skis that were six foot wide. Plus, the 300 horsepower engine that would be required is a bit heavy for most ski boats. This being said, I am always the first person asked to go along anytime family and friends go skiing. My job is to sit in the front of the boat and keep the bow down to help the driver pull the skiers. Around the lake, they affectionately refer to me as "ballast." They all say I'm much nicer to have along than the bags of cement they used to use.

A number of years ago my wife and I joined three other couples on a summer vacation to a popular Midwestern lake. We spent a good portion of that week out on the water. Each day I would take my spot in the front of the boat as people took their turns skiing and tubing.

One afternoon, my friend Dave bragged that he believed he couldn't be thrown off of the tube. Well, the boat captain accepted this challenge and told him to hang on. He first bounced Dave over the waves made by the boat and Dave just laughed and pretended that he was sleeping. So the captain put the boat in a tight spin. As Dave and the tube spun faster and faster it appeared to the rest of us that the tube was more or less airborne. Now Dave was actually six foot tall, but with the centrifugal force of the spin, he appeared to be about nine foot long. And I believe all the blood in his body was forced down to his feet. It looked like he had two giant watermelons attached to his legs. Dave finally gave up and let go. Now I'm sure most of you are like me and have at times tried to skip flat rocks over the surface of the water. We count the number of "rings" that the rock makes before it sinks. Eight or nine skips is pretty good. I'm not sure how many times Dave skipped, but I'd estimate it was about a hundred. By the time he came to a stop, he was in the next county. While boating over to pick him up we spied something floating in the water. It was Dave's swimming trunks. The swimming trunks were a brown color which was convenient given the circumstances. Dave was floating in his life jacket as we pulled up alongside him. The four women immediately peered over that side of the boat. Since the water was kind of cool, Dave was lucky that it was also kind of murky. We threw him a towel and dropped the ladder for him to climb aboard. Dave said the whole event wasn't that bad. He said he kinda got a kick out of it. After he let go and was skimming along the surface, he passed one of those big cigar boats out for a cruise. Dave gave a jaunty salute to the older, deeply tanned driver and the bikini-clad trophy wife out front. The driver was so miffed at being passed by a guy just skipping along that he took his cigar boat and headed back to the harbor.

That night while we were having drinks back at the cabin, Dave pulled me aside and asked if I had any aloe lotion. I laughed and said, "So, you got a little sunburned, did ya?"

"Actually," he replied, "I'm dealing with something like road rash on a part of my body that was supposed to be covered by my swimsuit."

I got him the aloe, but I didn't help him apply it.

Double G

CAST OF CHARACTERS

HERE'S SOME SHOCKING NEWS FOR you…. Times change. When we're gathered together nowadays, and our kids and grandkids are around, one of the little ones will do something cute or maybe run off with another one's toy and one of the women will smile and say, "Why that little Johnny…he's such a character." And for a moment I might agree, but then I think back to the stories I was told about some of the men in our little town from a few decades ago, and I think, "Now, there were some characters!" Back in the forties, fifties, sixties and even seventies, much of the entertainment in our community was provided by gentlemen who acted up in one way or another. I'm certain that it was just a coincidence that their actions and their patronage of the local taverns occurred at about the same time. Before there were televisions in everyone's home, men oftentimes frequented their favorite watering holes in the little towns like ours. In fact, in our village of fewer than 250 people, at least five different taverns were able to make a living at the same time. Guys just enjoyed socializing with their friends in the afternoon or evenings. It wasn't like drinking and driving was a problem since most everybody walked to their local bars. I was never told that these guys ever ran afoul of the law. In fact, there's a good chance the local constable was seated on the barstool next to them.

I was born in 1955. By the time I got old enough to appreciate some of the stories, some of these guys had passed away. But their stories lived on.

One guy frequented the same bar almost every evening. By the time 10 o'clock would roll around, he had usually managed to aggravate the bar owner to the point where the bar owner would throw him out. The gentleman would return the next evening and reseat himself at his favorite stool. After each beer he would yell at the bar owner, "How in my doing so far?" The bar owner would shake his finger at the guy. After about five or six beers…zoop! Out the door he'd go. He would yell back at the bar owner, "See you tomorrow night."

Another guy at that bar used to worry what his wife would say when he got home. So after every beer, he would walk behind the bar and lean in real close to the big mirror. When he could start to see red lines in his eyes, he went home.

I was with my dad one night at the one establishment when one of the regulars came out of the restroom on the far end of the building. He took off at a run heading straight towards the bar. At the last second, he slid across the floor, coming to a stop amid ten are more stunned patrons he looked up and asked, "Was I safe or was I out?"

Another guy used to come into one tavern and get a shot of "Red Eye," seven or eight times a day. When he was a middle-aged man, his doctor told him to cut back on the drinking, or he was going to die. He didn't cut back, and the doctor was right. Forty years later he died.

Not everyone who qualified as a character spent their time in a tavern. One old guy had the reputation of being one of the best winemakers anywhere around. He enjoyed partaking in his work. I had the opportunity to taste his wares long before I was legal drinking age. He told me it was all right if I used it for medicinal purposes. It must have worked because I don't remember coughing for a year.

One of the community's favorite characters was a milk hauler named Harry. Harry was a brother-in-law to my great aunt. I never saw him in the taverns, so I'm not aware that he drank much. In addition to hauling canned milk, Harry used to do odd jobs for people and every spring he plowed gardens for a lot of the town. Even though he had a small tractor and two bottom plow, Harry still did some of the field work with a team of horses. He was possibly the last person around who used horses to that extent. But Harry's best trait was his ability to tell stories. If there was any truth in his stories, it couldn't have been much. And even though we had heard his stories before, he told them with such enthusiasm and passion that you found yourself attentively waiting for the end. One of his favorites was the time he and a buddy went quail hunting. When a bird flushed in front of him Harry drew a bead on it and pulled the trigger. Just as he did he saw his buddy step out of the brush right in the line of fire. "Well," Harry said, "luckily my reflexes were so quick I was able to raise the gun barrel before the birdshot exited the gun."

Boy. I wish I had reflexes like that.

One character dear to me was my great uncle, Hemmy. Hemmy also hauled canned milk, and he had the ability to stretch a story as well. I wonder if that trait was passed on to any of our family? Two things stand out in my mind about Uncle Hemmy. The first one was his laugh. His laugh started out deep in his body and came out loud with a "Har, Har, Har , Har!" When Hemmy laughed you could hear him all over town. And he laughed a lot. He was a real genuine happy person. The second thing I remember about Hemmy was his "Red Man" chewing tobacco. Out of each pack he would get exactly two chaws. Half the pouch went in his mouth each time. If you didn't know he chewed tobacco, you would have sworn he had an infected tooth, and his mouth was filled with cotton balls. I may have a few of Uncle Hemmy's storytelling genes, but no offense Uncle Hemmy, I'm glad I didn't learn to chew tobacco. I've got so many other bad habits I don't have time for anymore.

One gentleman who stood out to me was Clem. He was one of my dad's best friends, and in my early years his 80-acre farm was just a half-mile behind our farm. Clem was a jolly guy, but he was also a man of few words. What I mean by that is, he could talk a lot, but he never believed in beating around the bush. When something needed to be said…he said it. When I was about 10 or 11 years old I was allowed to start rabbit hunting. I was borrowing my cousins 410 shotgun, and my dad took me to Clem's farm.

He had lots of rabbits in his pasture, so he and dad walked through the grass and tried to flush rabbits my way. After missing my third or fourth rabbit, Clem said, "Ain't much of a shot. Are you?"

"No. No, I'm not."

Then he and Dad laughed, and they found me a rabbit that didn't run as fast, and I actually hit it.

Clem was the kind of guy that if he didn't like the tie you were wearing, he wouldn't say, "I think that tie would look better with a different shirt." No, he would say, "That's a damn ugly tie."

"Yes. Yes, it is." Then he'd laugh.

As far as I know, Clem is the originator of one of the greatest lines I have ever heard. Some men were talking about when you knew that you had too much to drink and Clem said, "You're not really drunk until you're lying face down in the grass, holding on, afraid you're gonna fall off."

I've never heard it described any better.

After Clem quit farming and took a job in town, he occasionally worked with a guy who was a tad bit lazy. Clem would say of the guy, "He's just a little bit better than nothing."

Back when Clem was still on the farm and milking a few cows, Uncle Hemmy was his milk hauler. Hemmy also picked up the neighbors milk and that neighbor and Clem didn't get along. Hemmy had to drive a narrow little road between the two farms. One day Hemmy complained to Clem that there were sprouts growing out of the neighbor's pasture fence and they were scratching his truck. Clem said he would take care of it. That afternoon Clem chopped off the sprouts and left them in the neighbor's pasture. The next morning, the sprouts were in Clem's yard. Clem threw them back in the neighbor's pasture. Next morning they were back in Clem's yard. This went on for several months. Clem's neighbor must've gotten up in the middle of the night to accomplish this task because Clem never saw him. Either one could have just burned the sprouts, but neither one wanted to give in. I'm not sure what actually happened to the sprouts. They may have just rotted from age.

One time I saw that neighbor had his tractor and wagon stuck in his field while he was hand shucking corn. I took our tractor over to offer to pull him out. He promptly told me he didn't want my tractor putting big ruts in his field. Instead, he wanted me to dig trenches in front of his tractor tires while he shucked corn. I took about a half's second to think about it, then told him no. As I got on my tractor to leave I told him, "I hope you get out."

When I told that story to Clem sometime later, he said, "Well, (actually, I'm not sure that Clem used the word well, but, the word he used rhymed with well) I'd a told him I hope you never get out!"

And he would have too.

I suppose that we all had characters in our past that help to make us who we are today. Somehow I believe that the characters of today are a little more bleached and a lot less colorful. Maybe that's a good thing. But I have to tell you I'm very grateful to have known these guys, either in person or through their stories. Undoubtedly, there are many more people I could have added to this story. And I'm certain each of you can think of those who've influenced you in some way. I really believe that the lessons that we learn from the people around us are just as important as the lessons we learned while sitting behind a school desk.

When I drift off in a dream, and I see Clem standing next to St. Peter in heaven and St. Peter points down to the earth below and says, "Just look at some of those folks doing the work of the devil. How can they expect me to get them off of earth and into heaven?"

Clem would squint one eye, cocked his head a little to the side and say, "Well, I'd tell'em. I hope you never get out.

Double G

BOTTOMS UP

O NCE UPON A TIME THE weeds upon our land
  Really weren't that tough, we'd wipe them
   out by hand

We'd walk each planted row, with our
  old garden hoe

And chop the weeds all down and they'd no
  longer grow

They got a little stronger, when came a later date

And so we had to learn to go and cultivate

Two rows, six, then twelve, the cultivators grew

And the fields would look just fine by the time that
we were through

Then chemicals we got, to kill those nasty weeds

The products all were there, to satisfy our needs

Those weeds got even worse, and then they
did evolve

And how to rid them all, was difficult to solve

They seemed to laugh at sprays, they'd brush them
off with ease

The weeds were all but certain they had us
on our knees

There were some ugly weeds that seem to tolerate

All the sprays we'd try, at any given rate

But I am just too stubborn and I just won't give in

I've got a battle plan and I expect to win

The weeds upon my farm, well they are
not laughing now

I've opened up my shed and I'm bringing
out the plow

SPOTTY SHOWERS

A S I SAT AT MY desk and prepared to author yet another award-winning piece, my wife passed by and glanced over my shoulder at the title of my article.

"Spotty showers?" she inquired. "Is that a story of your infrequent attempts at personal hygiene?" As she walked away, her laughter was her own reward for her most recent witticism. Wives are so hilarious.

All the farmers know what I'm talking about. We've all experienced those rains that pop up on one end of the field and not on the other. Undoubtedly you've heard of the tales of how abruptly these rains can cut off. One guy I knew used to love to tell of the time he was driving a motorcycle while a friend was seated behind him and they were trying to outrun a thundershower. When they arrived at their destination, the driver was completely dry while his friend was soaking wet.

Another guy said he watched his dog trying to get home ahead of a storm. When the pooch finally slid to a stop on the front porch, his ears were dry while his tail had water running off it. Now, as you readers can obviously see, these tales are highly exaggerated by their tellers. I, on the other hand, only use factual truths when putting an article together. (Editor's note: "Okay folks, this is the part where you need to put your boots on!")

It happens every year that when you're harvesting at one farm, a downpour will occur while your farm a couple miles away stays dry…. Until you drive there with the combine.

Nowadays a lot of our trucks and wagons used for hauling grain have rollover tarps. A few years back, that wasn't the case. You could be waiting in line at the local grain elevator along with a dozen other trucks or wagons, when a surprise shower would pop up, sending everyone scrambling. Could you find a nearby shed are should you try to make it home? It was usually chaotic.

If you did happen to see the shower coming, you could call your wife and have her bring you a tarpaulin from your shed. Usually, the first raindrops were starting to fall by the time she got there. You quickly unfolded the tarp and tossed it towards the top of the truck bed just as a powerful gust of wind blew in. I've been told that the idea of parasailing came from watching farmers try to tarp their trucks. Most of the time I was lucky and landed back on the ground within a few blocks of the elevator. On the rare occasion that I came down in a neighboring state, I would just call for a taxi.

As farmers, we deal with a lot of people. One of the hardest to deal with is Mother Nature. Some of you might remember a commercial on television from a few years ago where Mother Nature is tricked into believing that what she tasted was butter while it was really margarine. She replies by saying, "It's not nice to fool Mother Nature!"

I learned this firsthand a few years ago. Sometimes it seems like mother nature is constantly working against me. When I need rain, it stays bright and sunny, and when I need dry weather, it pours. I had one field of beans left to plant, and they were beans that I was raising for a seed company. There was just enough bags to plant the field one time. I wouldn't be able to replant. The weather was nice enough, but I somehow knew Mother Nature was just waiting for me to plant these beans before she would unleash a torrent of rain. I didn't put the seed in the planter. Instead, I pulled the tractor and empty planter out to the field and proceeded to go up and down the field as if nothing was out of the ordinary. By the time I was on my last round, storm clouds were building in the distance. I just got the tractor in the shed when the skies opened up. It poured for almost an hour. Then the sun came back out, and a rainbow appeared in the East. Mother Nature was so happy with herself. I couldn't hold back my laughter anymore. I clapped my hands with glee as I came out of the shed, pointed up at the sky and said, "I got you Mother Nature! That planter was empty!" Like that, the rainbow disappeared, and a tornado carried my machine shed away. It really doesn't pay to try and fool Mother Nature.

Boy, talk about a touchy female.

To describe Mother Nature would be difficult. Usually, she's warm and sunny and nurturing. But occasionally, she can be wild and crazy. Actually that sounds a lot like this girl I used to date. She sure kept life interesting.

So I married her.

Double G

GO AHEAD.... MAKE MY DAY

I LIKE THE LIFESTYLE IN OUR little town

There is often a smile and seldom a frown

The houses are neat, and you know all your neighbors by name

There is a not as much crime as there is in the city

And the folks who live there, I can't help but pity

Cause they're missing out on some living, they might wish to claim

The pace of life here is simple you see

You can make it as hard as you wanted to be

and for exercise, there are those who just like to walk

You'd think that to be good and easy to do

But in my kind of town that's not always true

Because as soon as you're spied, you're expected to just stop and talk

So you halt, and you smile with your "how do you do's?"

And you inform one another on all of the news

That the local newspaper won't print till the following day

But you still buy the paper and read it all through

To confirm all the things that you already knew

And to learn of the yard sales that soon will be coming your way

Now the people work hard at the jobs that they choose

As they're toiling at work and paying their dues

To take care of a family that they hold so dear

Coming home to your town, well it helps you unwind

And the drive to your work you don't really mind

And the small-town advantage is always perfectly clear

You grab an iced tea and your old garden hoe

And you till up some soil where you think you will sow

Some kind of produce to save on your grocery bill

Now everyone knows that a garden is work

But fresh from the garden is kind of a perk

And to say "I grew this" you know is kind of a thrill

Now there's a church that sits on just down the street

Where a lot of the town will each weekend meet

When the steeple bells ring a beautiful welcome of love

The moms and the dads with the children in tow

And the grandparents too, they all want to go

To offer a praise and a thank you to the Master above

For the life that we live and the gifts we've been given

We know in our hearts are all sent from heaven

So will hear the good message and then we'll praise with a song

And we'll sit side-by-side with family and friends

When the amens shall come and the service doth ends

We'll offer best wishes till Sunday next comes along

Now there's towns just like this all over this land

And if one of them's yours well isn't it grand

So please stop on by if you're ever traveling our way

Cause if you're kind of country, I'm willing to bet

That you're one of my friends I haven't met yet

And making new friends, well that sure makes my day.

Double G

GARY ANDRETTI

NOW THAT I'M IN MY sixty's and my only active muscle is my stomach, I can accurately say that I'm not as fast as I used to be. I participated in all kinds of sports when I was younger, but I'd have to say my favorite was slow-pitch softball. I played on numerous teams over the first fifty years of my life, and I was never the fastest runner on any of those teams. But I was never the slowest either. I was always in the middle of the pack, and that was okay.

When it comes to my history of driving, you could probably say the same thing. I never tried to push my vehicle to its maximum speed. This may stem from the fact that the cars and trucks I drove weren't as new or fancy as my friends. If I went too fast, parts usually started falling off my vehicle, some of which were vital to keeping the thing running. One local mechanic would look at my car and say, "Don't go any faster than you want to crash."

Even today, I don't push the speed limit. My wife and I will be in the car heading to church or some other social event and of course, we are running late. She'll lean over from her passenger's seat to glance at the speedometer, then look at me and say, "I was just checking to see if you had it in gear."

Did I mention that she drives faster than me? She also runs faster which explains how she caught me when we got married. She's also a lot stronger than she looks. I remember right after I proposed she stopped twisting my arm behind my back and replied, "Oh my. This is also sudden. Let me think about it. O. K." As I recall, she got that whole reply in on one breath.

Now that you know a brief history about me, it might surprise you to know that I hold the fastest tractor speed ever recorded by our local police department. I say recorded, but that's not really true. The speed that allegedly showed up on the attending officer's speed gun was never actually verified by anyone else.

Now I realize that the souped-up pulling tractors would run very fast on a road. And the manufacturers are putting a lot faster speeds in the new model tractors to allow farmers less time traveling on the road. But neither of these factors figure into my run from the law.

The year was 1987, and I farmed land on two different sides of our little town. When you have a community that is surrounded by agriculture, it is not uncommon for tractors and implements to pass through its streets. The local police always treated the farmers very well, and I can say I knew most of the police force by name. The one officer told me their department considered it safer for me to pass through town on my tractor rather than my car. Although sometimes in late summer the police would call me up and asked me to bring my car to town and drive up and down the streets for a while. They said it saves them having to get the mosquito fogger out.

On the morning in question, I was heading through the town on my 1972 model tractor pulling a 12-foot wide offset disk. Unbeknownst to me, the town had just hired a new deputy on a trial basis. The officer was born and raised in an urban environment and had never encountered farm vehicles before. As I stopped at the town's only four-way stop, I noticed a squad car heading towards me from about three blocks ahead. I turn the corner and started through the downtown area. When I looked back, I heard and saw the squad car come flying through the four-way stop with his lights flashing and sirens blowing. I wondered what dastardly villain he was after. Turns out the dastardly villain was me. He motioned for me to pull over. Since I was close to the local filling station, I pulled my tractor and disk in there. I climbed down from my tractor while the officer got out of his car. His first words to me were, "I bet you're wondering why I stopped you?"

I replied, "Yah. I was kinda wondering."

"Well," he continued, as he pointed to the nearby thirty mile an hour speed limit sign, "that speed limit applies to tractors as well as cars."

Here's a bit of information some of you may not know. Laughing out loud generally doesn't improve the mood of the policeman.

The officer said, "I clocked your tractor doing 38 mph."

I tried to wipe the smile off my face as I told the man,

"No offense officer, but that tractor couldn't do 38 mph if you dropped it out of an airplane."

His expression led me to believe I had no future in comedy.

The officer told me he was letting me go with the verbal warning, but he added, "Maybe you can find another route to take instead of passing through town."

I promised to give it some thought and then got on my tractor and drove on. Now I know the young officer felt he was making the town safer for the residents, but I also know that he never had his radar gun pointed at me. The only time he saw me, I was stopped at the stop sign. Where the 38 mph figure came from, I'm not sure.

A few days later, I spoke with a different officer who was a good friend of mine. I asked him, "What's the deal with this new guy? He stopped me the other day and said I was going 38 mph on my tractor." I thought my friend was going to have a heart attack from laughing so hard. When he recovered he said, "Wait till I see him."

I found out later that the other officers constantly asked the new guy, "So, did you stop any speeding tractors lately?"

I'm pretty sure that by the time the guys trial period ended, he wouldn't have stopped a tractor even it if it was doing 50 mph.

At the end of his two weeks, the officer decided to take a job in Chicago, tractors being somewhat scarcer there.

And my claim to fame of having the fastest tractor around didn't last long. My neighbor put a V-eight engine into a Farmall M, and my officer friend went out to clock him, but he said his gun wouldn't go that high.

Double G

TRUST ME BABE...
I'M EXPERIENCED

IMAGINE YOU'VE HEARD THE PHRASE "Experience is the best teacher." I don't know. I'm afraid even with experience I would get a D minus. I do understand the idea though that by learning something hands-on, it makes it easier to remember. There's all sorts of things that I've learned around the farm through trial and error.... Mostly error…and probably an equal number of things that I'd just as soon forget. Like the fact that sheep can't swim. Or at least not all sheep. How that lesson was taught to us went like this. We usually kept a few head of sheep around the farm to help keep the grass down. More times than not they stayed in a little pen behind the grain bins. The pen had a little pond in it for the sheep to drink from. Late one summer after the sheep had all the grass clipped to the ground, we decided to move them to a different pen. We backed the cattle trailer up to the one gate and began herding the four head of sheep toward the trailer. Three out of the four sheep jumped on the trailer on the first try, but one old ewe decided she wasn't leaving. My brother and I chased her around the pond about a dozen times with her getting past us each time. We decided our best option was to just try and catch her and throw her on the trailer. So as our sons chased her towards us, we got ready for the tackle. When the woolly missile got close to us we did our best Dick Butkus imitation and leapt at her to bring her down. The old ewe jumped high over our outstretched arms and came down with a splash in the pond. Now the pond wasn't more than three or four foot deep, but as the wool of that old ewe started to absorb water she began to do <u>her</u> imitation of a submarine. Before she totally sank out of sight, my brother and I jumped in and started to pull her out. This ewe, which might have weighed a hundred and fifty pounds before, now felt like she weighed a thousand pounds. We grunted and groaned as we flopped around in the mud trying to get her back on the bank. Finally, the three of us were back on solid ground, and we found out we didn't have to worry about her running away because with the water in her wool she couldn't even stand up. It took ten minutes of squeegeeing her wool before she could get back on her feet. When we got her on the trailer, we were all exhausted. But it was a teaching experience. I learned to never buy sheep again. That's a lesson I still adhere to today.

One area I wish I had more experience in is electricity. I don't understand it all, so don't call me to wire your new house. When I need work done I call the professionals. My dad told me a story about a neighbor who gained some experience with electric fences back when they were new. Pete and Leo were cousins, and Pete was a few years older than Leo. Leo understood that the charge from an electric fence could pass through one person and into the next provided the next person was the end of the line so to speak. As they were walking next to an electric fence rabbit hunting one day, Leo would grab Pete's arm or shoulder with one hand then grabbed the hot wire with the other. Of course, Pete would feel the shock. After about two or three of the shocks, Pete was convinced Leo had electricity flowing through him. All Leo had to do was raise his hand and start towards him, and Pete would run.

Some learning experiences cost us money. The first combine I ever owned was an old 975 New Holland with a thirteen-foot grain platform and a gas motor. I didn't have great luck with that machine, and during the second year of harvesting soybeans, twice the bean fuzz got thick enough on the motor that it caught on fire. Both times I leapt from the cab and extinguished the flames. That winter when I wanted to trade the combine in, I was only offered about half as much money as I had the combine insured for. Experience says I shouldn't have put the fire out.

Here's another good lesson…. Listen to your father. Back when I was a young teenager and knew everything, I'm afraid I brushed off a little good advice from dad. I was in 4-H, and I spent the night at the fairgrounds the night before the 4H show. I had a large steer that I intended to show and dad warned me not to take the steer down to the wash rack the next morning before he got back to help me. But my youthful enthusiasm convinced myself I could handle it and I took the beast to the wash rack on my own. I got it washed and was leading it back to its stall when things went wrong. Just as I was passing by the north end of the exhibition ring, a beat-up car from the previous night's demolition derby passed by and backfired. That loud bang had the same effect on my steer that a starter's pistol would have on an Olympic sprinter. My steer headed for the south end of the arena determined to set a land speed record for the 40-yard dash. I was hanging on with both hands to the knot on the end of the lead rope. I had my heels dug in, but all I was doing was sending two plumes of sawdust into the air as I skied along behind the frightened animal. On the far end of the arena was a young man sound asleep on a cot. My steer was heading directly for him. I could see the headlines in the newspaper. "Sleeping man injured in skiing accident." I don't know if my steer got tired of dragging me or what, but he suddenly stopped with his head directly over the cot. I jumped up and ran to a nearby post and tied off the lead rope. When I looked back, the guy in the cot was still sound asleep. I've often wondered what that guy might have thought had he opened his eyes and saw this wild-eyed snorting beast just inches from his own face. I think that's the stuff nightmares are made of. I went and got my friend Rodney, and he helped me get my steer back to its stall. I was emotionally drained, and I knew I needed to confess to dad what I had done. And I did.

But by then, it was ten years later, and he could no longer outrun me. Confession may be good for your soul, but it's hard on your legs.

Another bit of advice my dad gave was to avoid women who were just after me for my chiseled good looks and my gobs of money.

I'm sorry. I'm going to have to stop writing now. I have to rush my wife to the hospital. She just leaned over my shoulder and read that previous paragraph, and now she can't stop laughing.

81

Double G

GRANDPA'S HOME

WELL, MOST OF THE STORIES and poems in this collection were written with the intent of getting a laugh or a smile. The following article is probably more directed at pulling up some memories for some of you readers. I don't need to tell you that too many folks, the land and the buildings associated with a farm are more than just assets. A part of your soul is embedded in everything that makes up the family farm. And many of you know the emotion that goes into the tough decision of selling the farm. The reasons that necessitate such a sale can be varied, but often, as in this case, there is no family member to take over, and the maintenance on the land and buildings becomes too burdensome for couples of retirement age plus.

This is the scenario that led to the farm that my son and daughter-in-law now call home. Perhaps it was fate that brought the selling family and our family together, but if you don't mind, I choose to believe it was divine intervention. Since our family met them, we have come to value the wonderful friendship of these people, and I can't imagine not knowing them as we do.

This selling couple has four beautiful daughters who all proved how intelligent they were by not marrying farmers like me. When the tough decision was made to sell the farm, a mutual friend introduced our two families. When we went to tour the farm, the dad, mom, and daughters guided us around the place, describing its features and relaying many memories that filled their minds. As they spoke, you could feel the emotion that spilled out of their hearts as well as their heads. This was not just a farm, this was a home.

My son and I looked at each other. There was no need to speak. We both knew that if at all possible, we wanted to be part of this special place.

Now there are farmers all over the Midwest that would have passed on the opportunity for this farm. The soil is not dark, the fields are not square, the yields are not record-breaking. There are woodlands and pastures and waterways and hills. But overall, the premises there is an aura of love that makes this place unique. Love by itself is not good collateral at the bank. But fortunately for us, we were able to secure the funds that made this transaction possible.

Throughout this time frame, I watched these wonderful people make some very tough decisions. My heart ached for them.

A few years have passed, and the plat book now shows

our family name on the land. But my son and I agree we will always refer to this property by the surname of the previous owners. We do this out of respect and acknowledgment as to what this farm became under their watchful eye.

While this was all taking place, I wrote a little song chronicling some of what they must have been feeling. The words of the song follow. I realize you can't hear the melody, but please imagine the emotion that is attached to these words. I hope you like the lyrics and for some of you it might invoke a memory. Here is:

GRANDPA'S HOME

Face-in goodbye is never easy
Never know just what to say
Three generations called this place home
Now we're movin away
The house we were born in was a place we could run to
Wherever we might roam
One thing you could count on
Was the love that was built in-to Grandpa's home

The farm was a haven us kids could all run to
When the world got to tough
Iced tea and a porch swing helped ease all of life's pain
If the goin was rough
The pasture and cornfields, the wind through the willows
Let you know that life goes on
And time could not weaken
The love that was built in-to Grandpa's home

The house may just look like lots of the others
In this world of push and shove
But this one is different cause it's held together
With God's grace and Grandpa's love

But sometimes things change, and we've had to move on
And our hearts have some doubt
But I think Gramps would like 'em, they're a God fear'in couple
And they're just startin' out
They've got lots to learn as they start down the pathway
To a family of their own

And they'll get lots of help
From the love that still lives in - Grandpa's home

Yah, Gramps would be smilin' because they're going to make it
In Grandpa's home